Bamboo Fences

Bamboo Fences

Isao Yoshikawa

With photographs by Osamu Suzuki

PRINCETON ARCHITECTURAL PRESS
NEW YORK

PUBLISHED IN 2009 BY
PRINCETON ARCHITECTURAL PRESS
37 EAST SEVENTH STREET
NEW YORK, NEW YORK 10003

FOR A FREE CATALOG OF BOOKS, CALL 1.800.722.6657.
VISIT OUR WEBSITE AT WWW.PAPRESS.COM.

FIRST PUBLISHED IN JAPAN UNDER THE TITLE
THE BAMBOO FENCES OF JAPAN IN 1988 BY
GRAPHIC-SHA PUBLISHING COMPANY LTD.
1-14-17 KUDAN-KITA
CHIYODAKU-TOKYO 102-0073
JAPAN

PHOTOGRAPHS: © OSAMU SUZUKI
TEXT AND DRAWINGS: © ISAO YOSHIKAWA

THIS EDITION © 2009 PRINCETON ARCHITECTURAL PRESS
ALL RIGHTS RESERVED
PRINTED AND BOUND IN CHINA
12 11 10 09 4 3 2 1 FIRST EDITION

FOR PRINCETON ARCHITECTURAL PRESS:
EDITOR: NICOLA BEDNAREK
DESIGNER: DEB WOOD
LAYOUT: BREE ANNE APPERLEY

SPECIAL THANKS TO: NETTIE ALJIAN, SARA BADER, DOROTHY BALL, JANET BEHNING, BECCA CASBON, CARINA CHA, PENNY (YUEN PIK) CHU, RUSSELL FERNANDEZ, PETE FITZPATRICK, WENDY FULLER, JAN HAUX, CLARE JACOBSON, AILEEN KWUN, NANCY EKLUND LATER, LINDA LEE, LAURIE MANFRA, JOHN MYERS, KATHARINE MYERS, LAUREN NELSON PACKARD, JENNIFER THOMPSON, ARNOUD VERHAEGHE, PAUL WAGNER, AND JOSEPH WESTON OF PRINCETON ARCHITECTURAL PRESS —KEVIN C. LIPPERT, PUBLISHER

YOSHIKAWA, ISAO, 1940–
[BAMBOO FENCES OF JAPAN]
BAMBOO FENCES / ISAO YOSHIKAWA ; WITH PHOTOGRAPHS BY OSAMU SUZUKI.
        p. cm.
ORIGINALLY PUBLISHED: THE BAMBOO FENCES OF JAPAN. TOKYO, JAPAN : GRAPHIC-SHA PUBLISHING CO., 1988.
ISBN 978-1-56898-834-4 (ALK. PAPER)
1. BAMBOO FENCES—JAPAN. I. SUZUKI, OSAMU. II. TITLE.
NA8390.5.Y67 2009
717—dc22
2008031589

EDITOR'S NOTE: WHERE POSSIBLE IN THE MAIN TEXT OF THE BOOK, FENCE NAMES AND TERMINOLOGY HAVE BEEN TRANSLATED INTO ENGLISH. THE JAPANESE FENCE NAMES AND TERMS FOR STRUCTURAL PARTS OF THE FENCE ARE GIVEN IN PARENTHESES AT THE FIRST MENTION.

Contents

PREFACE 06

BAMBOO FENCE STYLES
Kenninji Fence 08
Ginkakuji Fence 18
Shimizu Fence 22
Tokusa Fence 26
Teppō Fence 29
Nanzenji Fence 34
Spicebush Fence 36
 (*Kuromoji-Gaki*)
Bush Clover Fence 38
 (*Hagi-Gaki*)
Bamboo Branch Fence 40
 (*Takeho-Gaki*)
Katsura Fence 50
Bamboo Screen Fence 54
 (*Misu-Gaki*)
Raincoat Fence 58
Ōtsu Fence 62
Numazu Fence 66
Four-Eyed Fence 70
 (*Yotsume-Gaki*)
Kinkakuji Fence 80
Stockade Fence 88
 (*Yarai-Gaki*)
Ryōanji Fence 94

Kōetsu Fence 100
Nison'in Fence 104
Nanako Fence 106
Other Fences 108
Special Fences 112
Unique Fences 116
Wing Fence 120
 (*Sode-Gaki*)
Partitions, Gates, and Barriers
 Shiorido, Agesudo 132
 Niwakido 137
 Komayose 142
 Takesaku, Kekkai 146

BAMBOO FENCES
History 148
Materials 151
Classification 152
Glossary 156

Preface
Isao Yoshikawa

Bamboo, *take* in Japanese, is truly a versatile plant and has been put to practical use in numerous ways since ancient times. Thin bamboo poles have been inserted into the ground and interlaced with horizontal bamboo pieces to form a simple partition or a defensive barricade in virtually every country in which bamboo is found. There is even an ancient Chinese book entitled *Zhuzha* (Bamboo Fences). This word, written with the ideographs for "bamboo" and "stockade, fence," exists in both Chinese and Japanese; the Japanese pronunciation is *takesaku*. The combination of ideographs used to write another Japanese word for "bamboo fence," *takegaki*, seems not to exist in Chinese, however. While *takesaku* refers to bamboo structures used as simple partitions and defensive barricades, *takegaki* describes the beautiful, finely constructed fences, perhaps unique to Japan, that enliven the scenery of a garden. In the remainder of this essay, I use the term "bamboo fence" to indicate *takegaki*, the subject of this book.

While their construction has decreased in recent years, bamboo fences still exist in great variety in Japan. Since a given style may have several subtypes, there are well over a hundred different kinds of bamboo fences. How did such variation arise? One of the most important reasons is the development of the tea ceremony during Japan's Momoyama period (1573–1603) and subsequent developments in the design of tea ceremony gardens. The elegant bamboo fence, an important component of these gardens, came to be much beloved by adherents to the tea ceremony. There is an intimate relationship between the tea ceremony and Zen Buddhism, and visitors to Buddhist temples found certain partitions and enclosures used in Zen temples appealing and began to build them in tea ceremony gardens and the gardens of ordinary homes. Several fence styles are therefore named after Zen temples, such as the Kenninji fence, the Ginkakuji fence, and the Ryōanji fence.

The Japanese admire the freshness of new bamboo so much that until only a decade or so ago, they often rebuilt the bamboo fences of their gardens each year as part of their preparations for the New Year's celebrations. Even today, the traditional New Year's pine decoration, or *kadomatsu*, would not be complete without freshly cut bamboo. Yet the austere beauty of dried, brown bamboo also has much appeal, and because the lifetime of fresh bamboo is limited, a true bamboo fence is made of brown bamboo.

Another factor in the widespread development of bamboo fences was the large-scale cultivation of madake bamboo in Japan, the most suitable variety for fence construction. Madake bamboo is thin and perfectly straight, and the space between two successive joints is large, making it particularly good for horizontal frame poles and beading (the decorative molding along the top of a fence). Also called garadake bamboo, madake bamboo is the material of choice for the four-eyed fence (*yotsume-gaki*) and other fences. Another widely used variety is mōsōchiku bamboo. Although the trunk of this species is inferior to that of madake bamboo, its branches are pliant, so it is widely used in the making of bamboo branch fences (*takeho-gaki*).

It is fortunate for us that the photographer whose work appears in this volume, Osamu Suzuki, was so captivated by the beauty of bamboo fences. Here some 250 photographs from his collection, taken over many years, show the design of these bamboo fences, unique to Japan, in beautiful color. My contribution to this book is supplementary to Mr. Suzuki's. Using line drawings, I will attempt to explain some of my research on bamboo fences. I hope that readers will find the book useful and enjoyable.

Kenninji Fence

Employing many of the basic techniques of bamboo fence construction, the Kenninji fence is the most commonly made screening fence (*shahei-gaki*) in Japan. A fence at Kenninji, a Rinzai Zen temple in Kyoto founded in 1202, is said to be the origin of this type, although it is not certain. The arrangement of its horizontal and vertical elements give this fence a streamlined beauty.

There are three types of Kenninji fences: *shin*, *gyō*, and *sō*. This terminology was borrowed from calligraphy, where *shin* is the standard, straight style; *sō*, a freer cursive style; and *gyō*, a style between the two. The *shin* fence, which has beading (*tamabuchi*) at the top, is the most common. The number of horizontal support poles (*oshibuchi*) used for the Kenninji fence varies according to region.

Kansai-style Kenninji fence with four stout horizontal support poles

Seirakuji Temple, Fukuoka

KENNINJI FENCE

top Four-tiered Kenninji fence constructed around a maple trunk
Hakone, Kanagawa Pref.

bottom left Kenninji fence with five horizontal support poles of three-layered split bamboo
Ritsurin Park, Takamatsu

bottom right Kenninji fence with four horizontal support poles and widely spaced horizontal frame poles (*dōbuchi*)
Hōkyōin Temple, Kyoto

KENNINJI FENCE

top   Tall five-tiered Kenninji fence
Nijō Castle, Kyoto

bottom   Kenninji fence with double horizontal support poles and round-bamboo vertical pieces (*tateko*)
Meijō Park, Nagoya

KENNINJI FENCE

below Tall five-tiered Kenninji fence, used in place of a garden wall
Hama Rikyū Garden, Tokyo

top **Kenninji fence with vertical poles of garadake bamboo**
Hama Rikyū Garden, Tokyo

bottom left **Kenninji fence**
Hama Rikyū Garden, Tokyo

bottom right **The most common type of five-tiered Kantō-style Kenninji fence**
Inagi

KENNINJI FENCE

top   Drawing of Kenninji fence: *shin* style

bottom left   Small five-tiered Kenninji fence, a continuation of a wall
Sankōin Temple, Koganei

bottom right   Kenninji fence used as a partition within a garden
Kannonzen'in Temple, Musashino

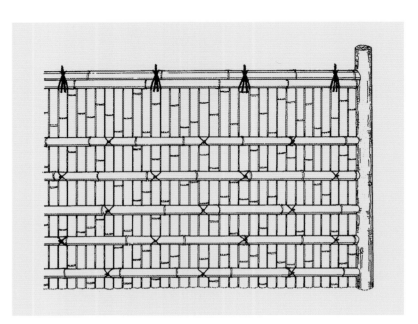

top left   Kenninji fence with tied decorations hanging from the beading
Niiza, Saitama, Pref.

top right   Kenninji fence with slender horizontal support poles, used as a blind
Tokyo

bottom right   Drawing of Kenninji fence: *gyō* style

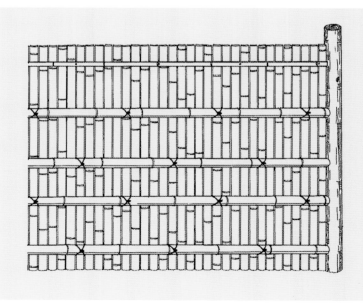

KENNINJI FENCE

below *Gyō*-style Kenninji fence without beading at the top
Ritsurin Park, Takamatsu

KENNINJI FENCE

top left — Kenninji fence with three stout horizontal support poles and logs at its base
Ryōanji Temple, Kyoto

bottom left — Kenninji fence with four-eyed fence at the bottom, an arrangement that is popular in the Kantō region
Rengeji Temple, Tokyo

right — Five-tiered Kenninji fence with wooden roof
Ritsurin Park, Takamatsu

Ginkakuji Fence

The Ginkakuji fence takes its name from the famous Zen temple in Kyoto, also known as Jishōji, built in 1474. The fence there tops a stone wall in the outer grounds on the approach to the temple from the main gate. Though there are some technical differences in construction, the Ginkakuji fence, usually made with two horizontal support poles, is very similar to a low Kenninji fence, and today low versions of the latter are generally called Ginkakuji fences. Resting on walls of stone or earth, this type of bamboo fence is perhaps the most beautiful of all.

The original Ginkakuji fence, made of madake bamboo

Jishōji Temple, Kyoto

GINKAKUJI FENCE

below A Ginkakuji fence that is almost identical to the original
Shōfukurō, Yōkaichi

GINKAKUJI FENCE

top Unusual Ginkakuji fence along a slope
Hōnen'in Temple, Kyoto

bottom The original Ginkakuji fence, atop a stone wall
Jishōji Temple, Kyoto

Shimizu Fence

The shimizu fence is constructed of shimizudake bamboo, a processed form of shino bamboo. Its structure is similar to that of the Kenninji fence, but with the thin shimizudake bamboo used for the vertical poles. The slenderness of shimizudake gives the fence a very beautiful appearance, but also makes maintenance rather difficult.

Fences made of reed, sarashidake bamboo (bamboo dried over a flame and oiled), or garadake bamboo are related to the shimizu fence because of their thin, round vertical poles.

Shimizu fence with vertical poles of slender sarashidake bamboo

Rengeji Temple, Tokyo

SHIMIZU FENCE

left Shimizu fence with vertical and horizontal poles of round bamboo
Keiō Hyakkaen, Tokyo

right Shimizu fence with vertical poles of slender shino bamboo
Kamakura, Kanagawa Pref.

SHIMIZU FENCE

top   Drawing of shimizu fence

bottom left   Shimizu fence with vertical poles made of reed
Kamakura

bottom right   A somewhat coarsely constructed shimizu fence of shimizudake bamboo
Kaizoji Temple, Kamakura

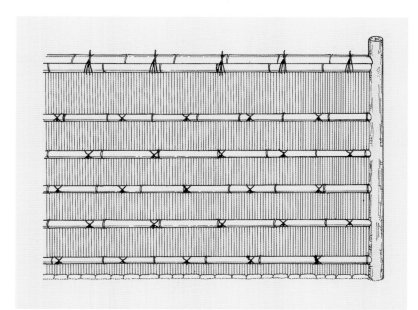

Tokusa Fence

Tokusa is a kind of rush. The tokusa fence is not in fact made of this plant, but is named after it because the vertical rows of bamboo resembles the rushes so often found growing in Japanese gardens. The plant's name is also found in the old term *tokusa-bari*, for walls of bamboo with this vertical arrangement.

The vertical poles are made of relatively thick madake bamboo split in half lengthwise. For ordinary walls, these would be fastened together with nails, but joining the bamboo with colored twine maintains the beauty of the tokusa fence. There are a variety of styles artisans use to tie the twine, making this aspect of the tokusa fence its most striking feature.

top Tokusa wall with changing heights separated by log posts
Sankōin Temple, Koganei

bottom left Tokusa wall with perfectly lined up split bamboo
Seirakuji Temple, Fukuoka

bottom right Tokusa wall exhibiting the beautiful color of dried bamboo
Kyoto

TOKUSA FENCE

top left   Large enclosing tokusa fence
Expo '70 Memorial Park, Osaka

top right   Tokusa fence between houses, with dyed twine
Tokyo

bottom left   Drawing of tokusa fence

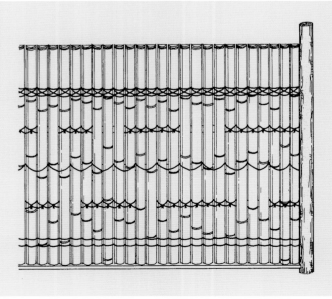

Teppō Fence

The term *teppō-zuke*, "with *teppō* (rifle barrels) attached," is a term in bamboo fence making that refers to a fence with vertical poles arranged alternately in front of and behind a horizontal frame. The vertical poles of most teppō fences are arranged in groups of a set number rather than alternating one by one.

Since the teppō fence is used mainly to prevent people on the outside from seeing in, it is classified as a screening fence. However, it can also be made into a see-through fence (*sukashi-gaki*), when single large bamboo poles are arranged alternately in front and back. Wing fences (*sode-gaki*) (see page 120) are commonly made in the teppō fence style as well.

Teppō fence by the entrance to a house, with vertical poles in groups of three and five

Kakueiji Temple, Yokohama

top left	Teppō fence with vertical poles of varying heights Kōetsuji Temple, Kyoto
bottom left	Teppō fence with paired horizontal frame poles of thin bamboo and vertical poles in groups of five Hama Rikyu Garden, Tokyo
top right	Teppō fence with two horizontal frame poles grouped together and vertical poles in groups of five Hakone, Kanagawa Pref.
bottom right	Teppō fence with front vertical poles in groups of three Jindai Botanical Garden, Tokyo

TEPPŌ FENCE

below Teppō fence with vertical poles of spicebush (*kuromoji*) branches bound together in the shape of torches (*taimatsu*)
Kyoto

TEPPŌ FENCE

top left — See-through teppō fence with vertical poles of irregular mōsōchiku bamboo
Tōkeiji Temple, Kamakura

bottom left — Teppō fence with vertical poles of stout mōsōchiku bamboo arranged one by one around the horizontal poles
Ryōanji Temple, Kyoto

top right — Very short teppō fence beside a pond
Yasukuni Shrine, Tokyo

bottom right — Drawing of teppō fence

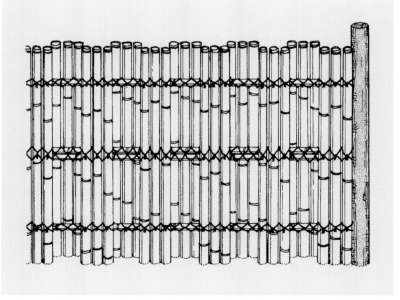

Nanzenji Fence

The Nanzenji fence takes its name from the fence in the back garden of the head priest's quarters at the Nanzenji temple in Kyoto (established in 1291), the headquarters of Rinzai Zen. This fence style is a mixture of the bush clover fence (see page 38) and Ōtsu fence (see page 62) styles. Most Nanzenji fences appear somewhat like a Kenninji fence, in which bamboo branches have been inserted, although the quintessential types resemble the original fence at the Nanzenji Temple, whose vertical pieces are assembled in the manner of the Ōtsu fence.

opposite top The original Nanzenji fence, with bush clover (*hagi*) branches distributed among the vertical split-bamboo poles
Nanzenji Temple, Kyoto

opposite left Nanzenji fence: a Kenninji fence with bamboo branches breaking up the vertical poles
Kōzōji Temple Betsuin, Machida

opposite right Nanzenji fence with sarashidake bamboo and bamboo branches
Kodaira, Tokyo

Spicebush Fence (*Kuromoji-Gaki*)

The term *kuromoji-gaki* is used for fences with frets (*kumiko*) made from the branches of the spicebush (*Lindera umbellata*). The spicebush fence is a type of brushwood fence (*shiba-gaki*), a class of fences made from the branches of various trees that predate the Heian period (794–1185). Because of the painstaking effort required for this process and the short supply of materials, the spicebush fence is one of the most expensive.

SPICEBUSH FENCE

left Five-tiered spicebush fence lining the path of a front garden
Katsura Imperial Villa, Kyoto

right Spicebush fence with roof and three horizontal support poles of stout split bamboo
Kyoto

Bush Clover Fence (*Hagi-Gaki*)

The bush clover fence is another type of brushwood fence, called *hagi-gaki* because its vertical pieces are made from branches of the *hagi*, or bush clover (*Lespedeza bicolor*). In contrast to the sturdy appearance of the spicebush fence, the thin branches of the bush clover give this fence a slender, delicate look, and it is used mainly for wing fences (see page 120).

A fence maker will often give the bush clover fence a rustic air by leaving off the beading at the top. This natural aspect can be further emphasized by purposely making the top of the fence uneven.

opposite top — Rustic clover bush fence with three horizontal support poles, used as a partition in a bamboo grove
Kairakuen Park, Milo

opposite left — Bush clover fence with four-eyed fence below
Jindai Botanical Garden, Tokyo

opposite right — Low bush clover fence along a corridor
Ryōanji Temple, Kyoto

Bamboo Branch Fence (*Takeho-Gaki*)

Takeho, which literally means "ear or head of bamboo," is a term used in fence making to refer to bamboo branches. Since many fences employ bamboo branches, the name is only used for fences built with this material if they do not have specific names of their own, such as the Katsura fence (see page 50) and the Daitokuji fence.

The bamboo used in these fences varies—some are made of stouter branches; others, of more delicate ones. The pliant branches of the mōsōchiku, hachiku, and kurochiku varieties of bamboo lend themselves well to use in these fences.

Bamboo branch fence with three horizontal support poles in a dry landscape garden

Nanzenji Temple, Kyoto

left Three-tiered bamboo branch fence
in the background of a garden
Tsurugaoka Hachimangu Shrine, Kamakura

right A pair of bamboo branch fences
Shinagawa Historical Museum, Tokyo

top   Four-tiered bamboo branch fence with bamboo branches wrapped around its vertical posts
Shōkadō Yawata, Kyoto Pref.

bottom   Four-tiered bamboo branch fence with fine bamboo *furedome* (see glossary, page 156) at the top
Koganei, Tokyo

top left	Six-tiered bamboo branch fence with closely spaced center horizontal poles Kamakura
bottom left	Four-tiered bamboo branch fence made of long branches Kamakura
top right	Black bamboo branch fence with five horizontal poles of thin bamboo and a fine split-bamboo *furedome* Rengeji Temple, Tokyo
bottom right	Bamboo branch fence flanking a gate Tōkeiji Temple, Kamakura

BAMBOO BRANCH FENCE

top — Long two-tiered bamboo branch fence in a bamboo grove
Sagano, Kyoto

bottom — Rustic bamboo branch fence along a garden path
Ōkōchi Sansō, Kyoto

top — Bamboo branch fence with twigs inserted diagonally between the upper horizontal support poles
Fukuoka

bottom left — Bi-level bamboo branch fence with bamboo joints beautifully arranged
Rakushisha, Kyoto

bottom right — Bamboo branch fence with diamond-shaped, see-through holes in the middle section
Yokohama

BAMBOO BRANCH FENCE

top Low bamboo branch fence with black bamboo
Kamakura

bottom Three-tiered bamboo branch fence
Kamakura

top left   Bamboo branch fence with coarse vertical bamboo branch poles
Hamamatsu Castle Park, Hamamatsu

bottom left   Combination Kenninji (bottom) and bamboo branch (top) fence
Ryōanji Temple, Kyoto

right   Drawing of bamboo branch fence

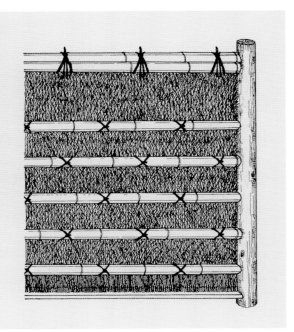

BAMBOO BRANCH FENCE

top   Special fence using bamboo branches as horizontal frets
Ōkōchi Sanso, Kyoto

bottom   Four-tiered bamboo branch fence with black branches
Kamakura

Katsura Fence

The Katsura fence takes its name from the fence surrounding the Katsura Imperial Villa in Kyoto, a famous park since the early seventeenth century. The fence was originally called *ho-gaki* (bamboo branch fence), because the name *katsura-gaki* was reserved for a hedge of black bamboo (hachiku) along the Katsura River in Kyoto. *Katsura-gaki* is now used throughout Japan to refer to this bamboo fence style, however, so this is the term we will use here.

The workmanship of the original Katsura fence is very intricate: rows of large and fine bamboo branches are arranged alternately to form a checkered pattern. Today, however, the fences are not usually made so elaborately.

The stately original Katsura fence

Katsura Imperial Villa, Kyoto

KATSURA FENCE

top The somewhat faded checkered pattern of the original Katsura fence
Katsura Imperial Villa, Kyoto

bottom left Vertical support poles emphasizing the beauty of the Katsura fence
Katsura Imperial Villa, Kyoto

bottom right Katsura fence of black bamboo branches in the background of a garden
Tsurugaoka Hachimangū Shrine, Kamakura

KATSURA FENCE

top The elegance of a low-lying Katsura fence in the winter
Irori-no-sato, Kodaira

bottom Detail of the above fence
Irori-no-sato, Kodaira

Bamboo Screen Fence (*Misu-Gaki*)

The *misu-gaki* is so named because it resembles the bamboo screens (*misu*) used inside the homes of the nobles of earlier centuries. The fences are also called *sudare-gaki*, as *sudare* is another word for these screens. Many bamboo screen fences are found in the Kantō region.

The most distinguishing feature of the construction of this fence is its fretwork: horizontal frets of sarashidake bamboo are set in grooves cut into the fence's posts. Vertical support poles are attached to the frets, giving the fence the appearance of a bamboo screen. Sarashidake does not hold up particularly well in the rain, but the light elegance of the fence makes up for this weakness.

Tile-roofed bamboo screen fence partitioning a garden

Tsurugaoka Hachimangu Shrine, Kamakura

top   Bamboo screen fence with paired vertical support poles of sarashidake bamboo
Koganei, Tokyo

bottom   Drawing of bamboo screen fence

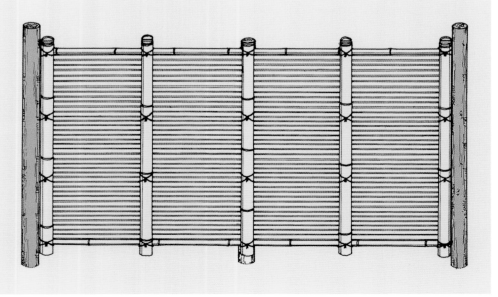

left Bamboo screen fence that steps up a slope
Tsurugaoka Hachimangū Shrine, Kamakura

right Bamboo screen fence with unevenly spaced vertical support poles
Tama, Tokyo

Raincoat Fence (*Mino-Gaki*)

The raincoat fence is one form of the bamboo branch fence; in earlier times it was also constructed with branches of the bush clover. The name *mino-gaki* is derived from the fence's resemblance to an old straw raincoat, or *mino*, because of the way the fine bamboo branches hang down. When it is constructed as a wing fence (see page 120), the raincoat fence is usually small. In the Kantō region, raincoat fences of kurochiku bamboo are popular.

The raincoat fence may be combined with other bamboo fences to make "half" raincoat fences (*han mino-gaki*), or the bottom may be left deliberately uneven to create a "broken" raincoat fence (*yabure mino-gaki*).

An unusual large raincoat fence atop a rock wall

Hakone, Kanagawa Pref.

RAINCOAT FENCE

top — Typical half raincoat fence with Kenninji fence at the bottom
Ōfuna Flower Center, Kanagawa Pref.

bottom left — Raincoat fence with see-through portion at the bottom and fine-bamboo horizontal support poles at the top
Kodaira

bottom right — Small raincoat fence beside a gate
Hakone, Kanagawa Pref.

RAINCOAT FENCE

top left — Half raincoat fence with four-eyed fence at the bottom and unique twine work
Tōyama Kinenkan Foundation, Saitama Pref.

bottom left — Raincoat fence constructed with large bundles of bamboo branches, perhaps a broken raincoat fence
Irori-no-sato, Kodaira

top right — Drawing of a raincoat fence

bottom right — Raincoat fence with thatched roof of bamboo branches and split-bamboo beading
Zushi

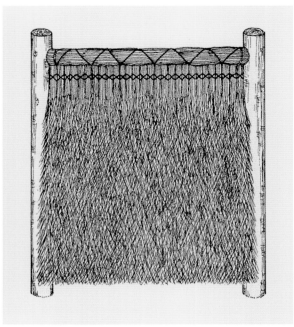

Ōtsu Fence

Certain very old varieties of bamboo fences were woven. The most iconic woven bamboo fence is the wickerwork fence (*ajiro-gaki*), and the most common wickerwork fence made today is the Ōtsu fence, also called *kumi kakine*.

The origin of the Ōtsu fence's name is not certain, although it is said to be derived from the fences that lined the highway passing through the city of Ōtsu in Ōmi Province (present-day Shiga Prefecture) during the Edo period (1615–1867). In the characteristic construction of the fence, several horizontal frame poles (*dōbuchi*) are attached to posts, and pieces of split bamboo or shino bamboo are woven into the poles.

Typical Ōtsu fence with horizontal frame poles of three pieces of split bamboo
Kyoto

top left   Ōtsu fence with some of the vertical poles facing the rear
Kyoto

top right   Ōtsu fence with three horizontal frame poles, each of two pieces of bamboo
Jindai Botanical Garden, Tokyo

bottom   Three-tiered Ōtsu fence with vertical poles inserted in between the weaves
Rinkyūji Temple, Kyoto

ŌTSU FENCE

top left — Bold Ōtsu fence with horizontal frame poles of stout round bamboo
Ryōanji Temple, Kyoto

top middle — Stylish Ōtsu fence with a horizontal support pole in the middle
Kyoto

top right — Detail of an Ōtsu fence with vertical poles of shino bamboo
Kamakura

bottom — Drawing of Ōtsu fence

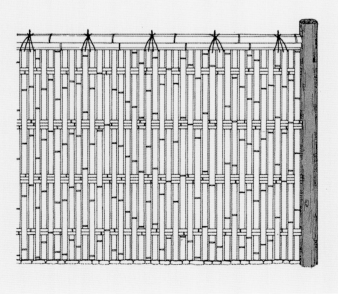

Numazu Fence

The Numazu fence is another variety of the wickerwork fence: its most prominent characteristic is the diagonal weave of its frets, which are usually made of slender shino bamboo. As a result of this weave, the front and the back of the fence are identical.

This style is called Numazu because the variety of shino bamboo most commonly used in its construction, hakonedake, is grown in the area around Numazu City, Shizuoka Prefecture. Finely split madake bamboo is also used to make such fences, although some people do not count these among the Numazu fences.

Numazu fence with diagonal weave of fine shino bamboo

Kōrakuen Park, Tokyo

below Tall Numazu fence with weave of split bamboo
Numazu

NUMAZU FENCE

left   Numazu fence of shino bamboo with horizontal support poles at the top
Wakayama Bokusui Kinenkan, Numazu

top right   Detail of a Numazu fence; the back surface of the split bamboo has turned black
Numazu

bottom right   Detail of a Numazu fence with split bamboo oriented and woven to create a pattern
Numazu

Four-Eyed Fence (*Yotsume-Gaki*)

The four-eyed fence is the most typical see-through fence and the most commonly constructed bamboo fence in Japan. Horizontal frame poles, usually four, are attached to the posts, and vertical poles are attached alternately in front of and behind the horizontal frame. Though this construction makes it a kind of teppō fence, the four vertical spaces resulting from the horizontal frame poles give this fence its name. Because of its simple form, making an interesting four-eyed fence can be difficult.

Like other kinds of bamboo fences, the four-eyed fence is made in three forms: *shin*, *gyō*, and *sō*. The fence is an important feature of tea-ceremony gardens, as it is generally found at the entrance to the inner garden, together with a gate of the same construction called *chūmon*.

Two examples of four-eyed fences
Shinagawa Historical Museum, Tokyo

FOUR-EYED FENCE

top   Unusual four-eyed fence with vertical poles of two different heights
Kamakura

bottom left   Two-tiered four-eyed fence used as a partition
Ikegami Baien, Tokyo

bottom right   Four-eyed fence, unusual in that short branches were left attached at the bamboo's joints
Sentō Imperial Palace, Kyoto

FOUR-EYED FENCE

left **Standard four-tiered four-eyed fence**
Kōrakuen Park, Tokyo

top right **Four-tiered four-eyed fence flanking a gate**
Katsura Imperial Villa, Kyoto

bottom right **Low-lying four-eyed fence serving as an inner partition in a tea-ceremony garden**
Seirakuji Temple, Fukuoka

FOUR-EYED FENCE

left Three-tiered four-eyed fence with two upper horizontal frame poles grouped together
Ritsurin Park, Takamatsu

right Three-tiered four-eyed fence with dark vertical poles
Ritsurin Park, Takamatsu

below Three-tiered four-eyed fences along a garden path
Ritsurin Park, Takamatsu

FOUR-EYED FENCE

top left — Low-lying four-eyed fence with three horizontal frame poles equally spaced
Kannonzen'in Temple, Musashino

bottom left — Winter view of a three-tiered four-eyed fence
Tonogayato Park, Kokubunji

top right — Three-tiered four-eyed fence with tall vertical poles tied together using the *karage* method (see glossary, page 156)
Shinagawa Historical Museum, Tokyo

top left   Simple *sō*-style four-eyed fence with vertical poles of varying lengths
Sōrōen, Tokyo

top right   Three-tiered four-eyed fence atop a stone wall
Kamakura

bottom right   Three-tiered four-eyed fence with a relatively wide space at the bottom
Rikugien Park, Tokyo

top Low-lying four-eyed fence with two narrowly spaced horizontal frame poles
Kōsokuji Temple, Kamakura

bottom Drawing of four-eyed fence

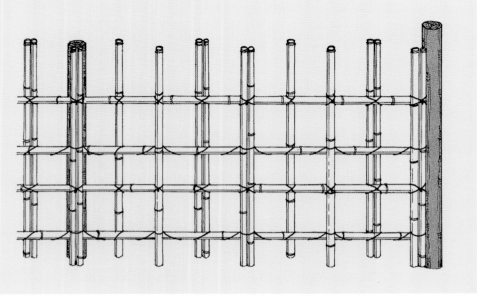

FOUR-EYED FENCE

top left   Four-eyed fence with only one horizontal frame pole
Kōrakuen Park, Tokyo

top right   Four-eyed fence with two widely spaced horizontal frame poles
Jindai Botanical Garden, Tokyo

bottom   Drawing of four-eyed fence with three horizontal support poles

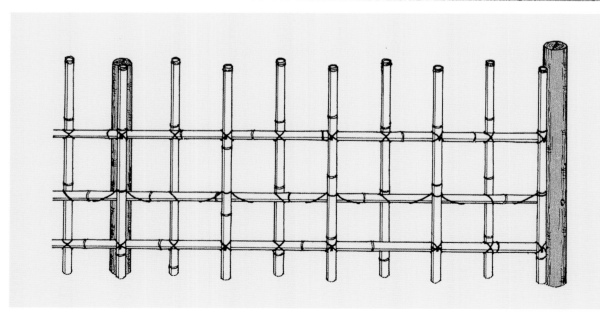

Kinkakuji Fence

Low-lying fences are called *ashimoto-gaki*, "foot-level fence." Of these, the Kinkakuji fence is the most renowned. The original Kinkakuji fence is in the northern section of the grounds of the Zen temple in Kyoto popularly called Kinkakuji, formally known as Rokuonji (built in 1397). This fence has both tall and short sections. The most prominent feature of a Kinkakuji fence is the split-bamboo beading along the top, and it is an attractive addition along a path in a front garden.

Kinkakuji fence in harmony with the stepping stones of a front garden

Keishun'in Temple, Kyoto

KINKAKUJI FENCE

top left — Kinkakuji fence with very close vertical poles
Chikurin Park, Kyoto

bottom left — Very tall Kinkakuji fence with two garadake bamboo horizontal poles and one of split bamboo at the base
Ikegami Baien, Tokyo

top right — Tall Kinkakuji fence with a split-bamboo horizontal support pole at the center
Hama Rikyū Garden, Tokyo

bottom right — Low Kinkakuji fence built on a slant along a stone stairway
Reiganji Temple, Kyoto

top Gate-front Kinkakuji fence
Jishōji Temple, Kyoto

bottom Unique Kinkakuji fence with three horizontal frame poles
Rengeji Temple, Tokyo

KINKAKUJI FENCE

below Kinkakuji fence with intermittent pairs of vertical poles
Kyoto

KINKAKUJI FENCE 85

top Atypical Kinkakuji fence, with stout-bamboo beading at the top
Tenryūji Temple, Kyoto
bottom Drawing of Kinkakuji Fence

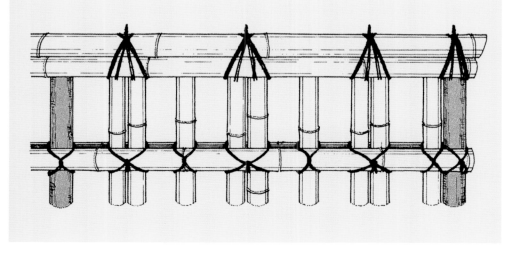

KINKAKUJI FENCE

top left — Tall Kinkakuji-style fence with an especially wide space between the beading and the horizontal support pole
Sōrōen, Tokyo

top right — Unique curved Kinkakuji fence
Meijō Park, Nagoya

bottom — Kinkakuji fence, very similar to the original, with two horizontal poles of garadake bamboo
Ikegami Baien, Tokyo

KINKAKUJI FENCE

top left — **Kinkakuji-style fence with horizontal frame poles and atypical vertical poles**
Kōrakuen Park, Tokyo

bottom left — **Bottom-heavy Kinkakuji fence with low split-bamboo horizontal poles**
Myōshinji Temple, Kyoto

top right — **Kinkakuji fence with paired vertical and horizontal support poles grouped together**
Kōsokuji Temple, Kamakura

bottom right — **Kinkakuji-style fence with four horizontal frame poles and garadake bamboo vertical poles**
Ueno Park, Tokyo

Stockade Fence (*Yarai-Gaki*)

The stockade fence has been widely constructed since the Edo period (1601–1867). *Yarai* is a word of broad meaning, referring to various barricades of wooden logs, which were formerly a common material for such. Bamboo was more inexpensive and practical than wood, however, so it became the material of choice for barricades, then called *takeyarai*.

In the most common style of construction, pieces of bamboo are sharpened at the tip to make frets, arranged crosswise diagonally, and attached to horizontal frame poles. A low-lying style in which heavy pieces of bamboo are used for the frets is common in the Kansai region.

Stockade fence of fine construction surrounding a large tree
Tsurugaoka Hachimangu Shrine, Kamakura

top left   Back of a long two-tiered stockade fence along a garden path
Ritsurin Park, Takamatsu

bottom   Drawing of stockade fence

top right   Three-tiered stockade fence with diagonal frets at an acute angle
Tsurugaoka Hachimangō shrine, Kamakura

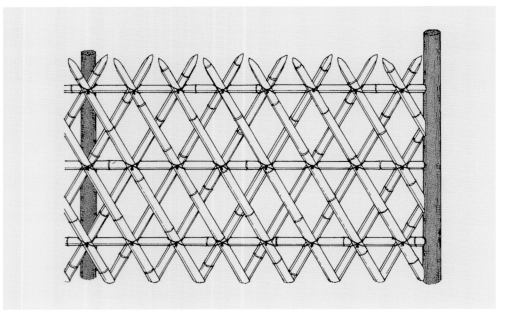

STOCKADE FENCE 91

top   Two-tiered stockade fence with tops of the frets cut horizontally
Meijō Park, Nagoya

bottom   Two-tiered stockade fence, unusual because the upper horizontal support pole does not intersect the frets where they cross
Kairakuen Park, Mito

STOCKADE FENCE

top left   Unusually large four-tiered stockade fence
Kairakuen Park, Mito

bottom   Stockade fence with exposed cut ends of garadake bamboo
Rikugien Park, Tokyo

top right   Front of a two-tiered stockade fence built on sloping ground
Ritsurin Park, Takamatsu

STOCKADE FENCE

left Three-tiered stockade fence with frets set at a wider angle
Hamamatsu Castle Park

right Three-tiered Kansai-style stockade fence in a bamboo grove
Kairakuen Park, Mito

Ryōanji Fence

The Ryōanji fence is a fence of superior construction. It resembles a stockade fence with beading at the top, which is its most prominent feature. Like the Kinkakuji fence, it is low lying. The original fence of this name, which lines the main path on the grounds of Ryōanji, a Zen temple in Kyoto, is constructed of frets of double-layered split bamboo.

Ryōanji fences of fine unsplit bamboo also exist, however. The Ryōanji fence has recently found widespread use in small gardens, while examples of the large-scale true style have become rare.

The original Ryoanji fence, with frets that are not inserted into the ground

Ryoanji Temple, Kyoto

RYŌANJI FENCE

top   Ryōanji fence with horizontal support poles grazing the ground
Ritsurin Park, Takamatsu

bottom   Drawing of Ryōanji fence

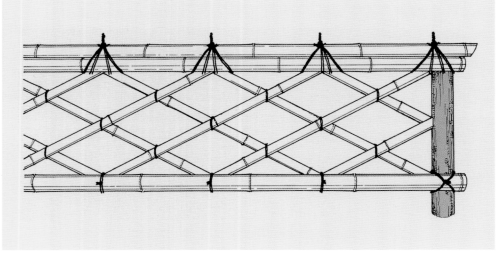

top Especially tall Ryōanji fence with frets inserted into the ground
Meigetsuin Temple, Kamakura

bottom A corner of the Ryōanji fence with horizontal support poles grazing the ground
Ritsurin Park, Takamatsu

top   **Tall, winding Ryōanji fence with especially fine frets**
Seiryōji Temple, Kyoto

bottom   **Ryōanji-style fence with frets at a very acute angle**
Seiryōji Temple, Kyoto

top left	Ryōanji-style fence with beading at the top of a stockade fence Seirakuji Keiō Hyakkaen, Tokyo	*top right*	Protective Ryōanji fence surrounding pine trees Kairakuen Park, Mito
bottom left	Small Ryōanji fence with horizontal support poles and beading of bush clover branches Shōkadō, Yawata, Kyoto Pref.	*bottom right*	Ryōanji fence with two horizontal support poles, very similar in form to a stockade fence Kairakuen Park, Mito

Kōetsu Fence

Hon'ami Kōetsu (1556–1677) was a craftsman of many talents who lived in the Takagamine district of Kyoto. He constructed the original Kōetsu fence on the grounds of his family temple, Kōetsuji, separating a tea garden and a hermitage.

Also called *kōetsuji-gaki* (Kōetsuji fence) or *gagyū-gaki* ("prostrate cow" fence), the Kōetsu fence is a see-through fence and similar in construction to the stockade fence. It is characterized by round beading at the top wound with bamboo branches and split bamboo, and by the way its end curves down to the ground. The Kōetsu fence is also used as a short wing fence.

The large original Kōetsu fence, whose pillars and beading are wound with split bamboo

Kōetsuji Temple, Kyoto

KŌETSU FENCE

top — Drawing of Kōetsu fence

middle left — Kōetsu fence, the main feature of this dry landscape garden
Shōfukurō, Yokaichi, Shiga Pref.

bottom left — Detail of the original Kōetsu fence
Kōetsuji Temple, Kyoto

bottom right — Small Kōetsu fence partitioning a garden
Kōzōji Temple, Machida

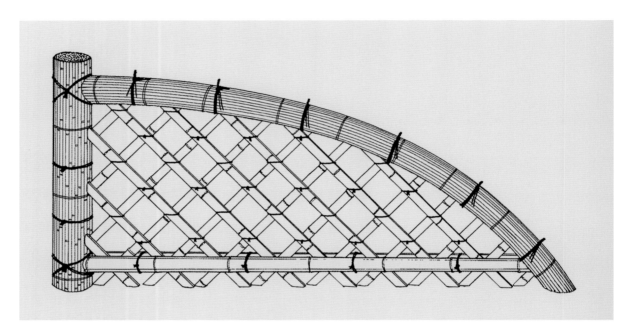

top left   Kōetsu-style fence with a bamboo screen attached to the frets on the right
Ikegami Baien, Tokyo

bottom left   Roofed Kōetsu fence with bamboo-branch posts and beading
Kyoto

top right   Low-lying Kōetsu fence with frets and beading made entirely of bush clover
Nijō Castle, Kyoto

bottom right   Kōetsu fence with stout beading and especially fine frets
Chikurin Park, Kyoto

Nison'in Fence

Nison'in is a temple of the Tendai sect of Buddhism in Saga, Kyoto. This unique low-lying fence, sometimes stretching in a straight line, sometimes curved, lies inside its grounds, enclosing the front garden of the main hall. It is very unusual for a fence to be named after a temple that belongs to a sect other than Zen.

The Nison'in fence is a relatively new style of bamboo fence, and it is similar in form to the Kinkakuji fence. A diagonal fret is inserted between each of the vertical pieces.

top left   Circle-shaped Nison'in fence with diagonal frets as reinforcement
Nison'in Temple, Kyoto

top right   Straight Nison'in fence
Nison'in Temple, Kyoto

bottom   The original Nison'in fence, a beautiful low-lying partition within the temple grounds
Nison'in Temple, Kyoto

Nanako Fence

The word *nanako* refers to a twill weave, and it is thought that this fence's name comes from an old weaving style. It is made by curving finely shaved pieces of bamboo and inserting them into the ground, making this the simplest form of bamboo fence. When the bamboo pieces are inserted into a base, nanako fences are movable. These fences are usually used in public parks to keep people out of certain areas, rather than in the gardens of homes.

top left Nanako fence with a horizontal frame pole of double-layered fine split bamboo
Ritsurin Park, Takamatsu

bottom left Slanted nanako fence along a garden path
Kōrakuen, Tokyo

right Movable nanako fence
Urasenke, Kyoto

Other Fences

Additional fence types are described on the following pages through photographs. One more fence, however, deserves some mention in detail.

The torch fence (*taimatsu-gaki*) is, broadly speaking, a bamboo fence with vertical pieces made of bush clover or spicebush branches bundled in the shape of torches. To avoid confusing this style with the teppō fence, which is sometimes constructed with bundled branches, it is best to only call a fence a torch fence when the bundles are attached to only one side of the horizontal frame poles.

Torch fence with spicebush vertical poles
Kyoto

top left   Fence with large bundles of bamboo branches shaped like torches

Chikurin Park, Kyoto

bottom left   Fusuma fence of shino bamboo flanking a gate

Hama Rikyū Garden, Tokyo

top right   Armor fence (*yoroi-gaki*) made of three stacked raincoat fences

Kamakura

bottom right   Tachiai fence with alternating pairs of bamboo and bundled vertical poles

Kairakuen Park, Mito

OTHER FENCES

top left   Sharp spikes atop a wall
Kyoto

bottom left   Folding-fan fence (ōgi-gake), with only slight openings
Nanzenji Temple, Kyoto

top right   Hishigi fence atop a stone wall
Kyoto

bottom right   Simple fence, perhaps a variation of the nanako fence, along a stone path
Tōkeiji Temple, Kamakura

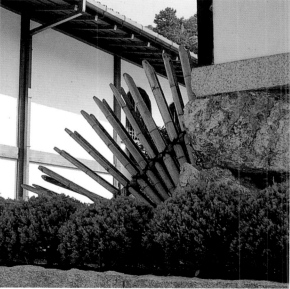

Special Fences

In this book, the term "special fences" refers to fences not readily classifiable in any of the categories mentioned previously. Although some special fences do have specific names in different parts of the country, some of these names are used only in the trade and are not widely known.

SPECIAL FENCES 113

top left   Garadake bamboo fence
Hama Rikyū Garden, Tokyo

bottom left   Fence with horizontal support poles and fine-bamboo vertical poles of varying heights
Ikegami Baien, Tokyo

right   Curved fence with vertical poles of garadake bamboo
Ikegami Baien, Tokyo

top   Movable fence with two horizontal support poles beside a pond
Ikegami Baien, Tokyo

bottom   Four-eyed-style fence with bamboo branches inserted at the top and bottom
Kawasaki

SPECIAL FENCES

left   Screening fence whose vertical poles of round bamboo have been cut off diagonally at the top
Hakone Art Museum, Kanagawa Pref.

top right   Katsura-style fence with horizontal split-bamboo frets
Tokyo

bottom right   See-through fence with coarsely woven split bamboo
Jōshōji Temple, Kyoto

Unique Fences

Unique fences have been designed by a landscape architect to fit a particular garden. Most of these are constructed by adding a modern touch to a traditional bamboo fence style. The most common model is the Kenninji fence; its vertical pieces can be rearranged horizontally or diagonally and attached to support poles. Designers often give their creations an original name. Two well-known unique fences are the *aboshi-gaki* (net-drying fence) and the *moji-gaki* (letter fence).

A bold unique fence with a design reminiscent of swirls of clouds

Ryōgin'an, Kyoto

below A unique fence modeled after bolts of lightning
Ryōgin'an, Kyoto

UNIQUE FENCES   119

left A unique fence with diagonally arranged frets
Seirakuji, Fukuoka

right A unique fence in which branches have been left on the bamboo
Ryōgin'an, Kyoto

Wing Fence (*Sode-Gaki*)

A wing fence is a fence with an end post inserted into the ground next to a building; the fence extends out from this post in the shape of a kimono's sleeve. Most wing fences, serving both for stylish decorations and for privacy, are very exquisitely made. They are usually screening fences, but because of the unique design of many wing fences, they are presented as a separate category in this book.

Kenninji wing fence with different-sized spaces between the horizontal support poles

Keiō Hyakkaen, Tokyo

WING FENCE

left Low wing fence with vertical poles and horizontal support poles of kurochiku bamboo
Nezu Art Museum, Tokyo

right Teppō wing fence with horizontal frame poles, posts, and vertical poles of bush clover
Shōkadō, Yawata, Kyoto

left — Five-tiered shimizu wing fence with vertical poles of shimizudake bamboo
Hakone Art Museum, Kanagawa Pref.

top middle — Torch wing fence with vertical poles of bush clover
Jōshōji Temple, Kyoto

bottom middle — Teppō wing fence with four horizontal frame poles and stout round-bamboo vertical poles
Tonogayato Park, Kokubunji

right — Bamboo branch wing fence with somewhat coarse branches
Rakushisha, Kyoto

top left	Small wing fence with coarse vertical poles of spicebush branches Keishun'in Temple, Kyoto	*top right*	An elegant tea-garden wing fence with vertical posts wrapped with fine split bamboo Kōetsuji Temple, Kyoto
bottom left	Small wing fence with a stout post and beading of bundled spicebush branches Jōshōji Temple, Kyoto	*bottom right*	Tall wing fence similar to that on page 125, top left, but with vertical posts of bush clover Kyoto

top left Spicebush wing fence with diamond-shaped openings at the top
Rakushisha, Kyoto

top right Unusual six-tiered bamboo branch wing fence with split-bamboo pieces attached vertically to the horizontal support poles
Ōkōchi Sansō, Kyoto

bottom right Wing fence made with small bamboo branches
Ikegami Baien, Tokyo

left   Bamboo branch wing fence made of mostly black bamboo, with some ordinary bamboo interspersed
Tokyo

top right   Five-tiered bamboo branch wing fence with stout bundled beading
Kakueiji, Yokohama

bottom right   Typical Kantō-style bamboo branch wing fence
Tonogayato Park, Kokubunji

below Somewhat small, but refined, Katsura wing fence and *shiorido*
Rakushisha, Kyoto

top left   **Bamboo screen wing fence with frets of fine shino bamboo**
Kyoto

bottom left   **Half raincoat wing fence: raincoat fence at the top and four-eyed fence at the bottom**
The Tōyama Kinenkan Foundation, Saitama Pref.

top right   **Bamboo screen–style fence with frets protruding from both sides**
Jindai Botanical Garden, Tokyo

bottom right   **Broken raincoat wing fence**
Shinagawa Historical Museum, Tokyo

top left   An especially coarse raincoat wing fence of black bamboo branches
Koganei, Tokyo

bottom left   Four-tiered wing fence at a tea-ceremony house, with vertical pieces of kurochiku bamboo
Keiō Hyakkaen, Tokyo

right   Wing fence with diamond-shaped holes at the top and a raincoat fence at the bottom
Ikegami Baien, Tokyo

top left Wing fence with fine bamboo vertical poles attached to wooden horizontal frame poles
Nijō Castle, Kyoto

bottom left Simple wing fence with a frame of plum wood and garadake bamboo frets
Sankōin Temple, Koganei

right Refined wing fence with split bamboo arranged vertically in a fish-scale pattern
Irori-no-sato, Kodaira

left   Delicate wing fence with very fine frets
Numazu

right   Small wing fence attached to the pillar of a garden gate
Kōsokuji Temple, Kamakura

Partitions, Gates, and Barriers
Shiorido, Agesudo

A *shiorido* is a very simple, light partition inside a garden made by wrapping thin strips of bamboo cladding around a rectangular frame of fine round bamboo poles, weaving the strips into a diamond-shaped pattern. The name *shiorido*, or "bent-branch door," comes from the way the strips of bamboo cladding are bent around the frame.

An *agesudo* is a *shiorido* suspended from a frame; it can be raised with a bamboo pole to let people in and out. Both of these gates are popular in tea-ceremony gardens.

Freshly built *shiorido* at the entrance to a tea-ceremony garden

Kohōen, Tochigi Pref.

PARTITIONS, GATES, AND BARRIERS

top left — Very low *shiorido* for partitioning and decoration
Kohōen, Tochigi Pref.

bottom — Low *shiorido* that beautifully complements the autumn dōdan azaleas
Kohōen, Tochigi Pref.

top right — Standard-sized *shiorido* with a single thin piece of bamboo at the top
Meiji Park, Nagoya

top left   A slightly damaged unusual *shiorido* with inner vertical frame poles in addition to the regular frame
Nezu Art Museum, Tokyo

bottom left   Tall double-doored *shiorido*
Ōkōchi Sansō, Kyoto

top right   Uniquely made *shiorido* woven with reeds
Kyoto

bottom right   Well-built *shiorido* for the inner gate of a tea-ceremony garden
Shinagawa Historical Museum, Tokyo

top   An *agesudo* suspended from a large *shiorido*
Kohōen, Tochigi Pref.

bottom   The same *agesudo*, raised with a bamboo pole to let guests in
Hakone, Kohōen, Tochigi Pref.

Niwakido

A *niwakido* is a wooden gate, sometimes with a small roof, found between the front garden at the entrance of a home and the main garden. Although the *niwakido* is not a bamboo fence, a few examples are presented here because the doorway is often elegantly constructed from bamboo.

Simple *niwakido* with wooden posts and paired vertical posts

Meijō Castle, Nagoya

below ***Niwakido*** featuring vertical pieces of garadake bamboo and a cypress-bark roof
Kohōen, Tochigi Pref.

top left   Beautiful *niwakido* with narrowly spaced vertical pieces of sarashidake bamboo
Shinagawa Historical Museum, Tokyo

bottom left   Wide *niwakido* with split bamboo in the tokusa style
Koganei

right   *Niwakido* with a door featuring vertical pieces of especially fine bamboo
Tsurugaoka Hachiman Shrine, Kamakura

top left   Winter view of a *niwakido* with fine bamboo arranged in diamond patterns within the wooden frame
Tonogayato Park, Kokubunji

bottom left   Niwakido with bamboo roof and diamond-patterned door made of frets
Ikegami Baien, Tokyo

right   *Niwakido* with fine split bamboo in diamond shapes at the top and latticework at the bottom
Ritsurin Park, Takamatsu

Komayose

The word *komayose* (or *komayoke*), which refers to a low barrier placed in front of a house to prevent horses (*koma*) from entering, is not used much today. In Kyoto one can find a type of *komayose* made of latticed bamboo that is placed at an angle against the lower part of a house's wall.

Modern *komayose* are no longer needed to keep horses out, so they have taken on a more decorative purpose. Still, they are installed to protect against splashing rain, mischievous children, and so on.

Komayose with split bamboo carefully fastened to the four horizontal frame poles; note the distribution of the bamboo joints

Kyoto

top left   **Five-tiered *komayose***
Kyoto

bottom left   **Standard four-tiered *komayose***
Kyoto

right   **Curved *komayose***
Kyoto

top left	Three-tiered *komayose* set at an acute angle on masonry Kyoto
bottom left	*Komayose* with garadake bamboo horizontal frame poles and supported by logs Kyoto
top right	Low-lying *komayose* of paired, split bamboo on alternating levels Kyoto
bottom right	*Komayose* with vertical poles leaning slightly outward Tokyo

Takesaku, Kekkai

As stated in the preface, *takesaku* ("bamboo fence"), as it is used here, refers to simple barriers and not to the relatively complex structures covered by the term *takegaki*.

Kekkai is a word from Buddhism that originally meant a fence or wall separating a place from the outside world, for ascetic training. The term is also used to refer to simple barriers to prevent people from entering a garden.

opposite top — *Takesaku* with two horizontal frame poles
Kohōen, Tochigi Pref.

opposite left — *Takesaku* with stout bamboo on log posts
Kōtōin Temple, Kyoto

opposite middle — *Takesaku* with garadake bamboo inserted into stout bamboo posts
Hōnen'in Temple, Kyoto

opposite right — Winter view of a *kekkai* with stout bamboo on log posts
Tonogayato Park, Kokubunji

Bamboo Fences
Isao Yoshikawa

Fig. 1

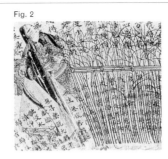
Fig. 2

HISTORY

Species of bamboo are found throughout the world. If the varieties of bamboo grass (*sasa*) are excluded from the classification, however, we find that most of the practically useful, high-quality bamboo is found in Asia. Bamboo is a symbol of the cultures of the East, especially those of China and Japan.

China has a long history of using bamboo. Bamboo was being made into various practical articles even before the Common Era, and images of the plant were incorporated into paintings and poetry from early times. An old stone monument carved with a beautiful representation of bamboo at the famous Han-dynasty Qufu Confucian shrine has come to be known as the "bamboo leaf stone." In the literary field, such ancients as Bo Juyi of the Tang dynasty and Su Dongpo of the Northern Sung dynasty employed images of bamboo in their writings. One early and well-known motif is the theme of "the seven wise men of the bamboo grove," a group of Taoists allegedly endowed with supernatural powers.

Looking at such a rich history, one would think that bamboo fences, or *takegaki*, would have been widely developed in China, but my research suggests that this was not the case. Simple bamboo barricades existed from early times, but more intricate fences such as those in Japan were hardly ever constructed. One exception are the fences made to enclose flowers. Numerous Ming-dynasty paintings depict these fences, which are thought to have first been made during the Tang dynasty. Chinese bamboo fences reminiscent of Japanese *takegaki* (Fig. 1) can be found in ancient Chinese gardens preserved to the present day, but they are very few in number and do not exhibit the great variety of bamboo fences found in Japan.

It is known that bamboo was appreciated from early times in Japan because of references to it in *Manyōshū*, a collection of poems from the eighth century. By the Heian period (794–1185), bamboo had found its way into the writings of many poets. Bamboo was at that time a symbol of coolness and so, according to old records, was planted on the north side of nobles' residences.

Although the construction of bamboo fences had not reached the state of development evident today, numerous forerunners of the *takegaki* were made during Heian times. The prime example is the brushwood fence, made by clumping together branches of various trees, arranging them vertically, and holding them together with horizontal support poles made of the same clumped branches. These fences give their surroundings the air of a mountain hamlet and thus often appear in the literature by women writers of that era as representations of *mono no aware*, "the pathos of things." In *The Tale of Genji*, a classic work of Japanese literature attributed to the noblewoman Murasaki Shikibu (early eleventh century), we have a prime example: "Surrounded by a rustic brushwood fence was a garden scrupulously planted." The scroll of *The Tale of Genji* and other paintings of the Heian and Kamakura (1185–1333) periods have left us with many depictions of what brushwood fences looked like then (Fig. 2). At Nomiya Shrine in Sagano, Kyoto, there remains a simple but attractive small brushwood fence from these

Fig. 3

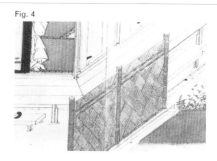

Fig. 4

early times (Fig. 3). Because the brushwood fence is considered the direct ancestor of fences made of bamboo branches, it is today classified as a bamboo fence, even though it is not made of bamboo.

Other Heian-period fences mentioned in early writings are the *suigai*, the *tatejitomi*, and the *higaki*. The word *suigai* is derived from the same elements as the word *sukashi-gaki*, the general term for see-through bamboo fences. *Suigai* were partitions of fine slats of wood or bamboo woven together loosely, so that one could see through them. In *The Tale of Genji* we read: "A bamboo *suigai* enclosed the princesses' room, forming a particularly severe boundary."

Tatejitomi were built as partitions within a room. A passage from the *Pillow Book* (around 1001 to 1010) by the Japanese noblewoman Sei Shōnagon gives evidence that they were different from *suigai*: "When one sits near a *tatejitomi* or *suigai*, there is a certain elegance in hearing a voice saying, 'it looks like it's going to rain.'" Although most were made of Japanese cypress bark (*hiwada*) and woven in wickerwork fashion, some *tatejitomi* were made of bamboo. In a fourteenth-century picture book depicting the Buddhist monk Kōbō Dai-shi's teachings (*Kōbō Daishi Gyōjō Emaki*), a *tatejitomi* with end posts of stout bamboo is illustrated (Fig. 4).

The *higaki*, or Japanese-cypress fence, is a partition of thin slats of Japanese cypress woven in wickerwork fashion and is therefore very closely related to the *tatejitomi*. The *higaki* is defined in *Wakun no Shiori*, an Edo-period (1615–1867) dictionary, and the eleventh-century book of tales *Konjaku Monogatari* speaks of "a very large house enclosed by a very long *higaki*."

In any event, the eventual development of bamboo fences came about as a result of successive gradual changes in these early partitions. Fences similar to what we know today as bamboo branch fences and Kenninji fences began to appear during the Kamakura period, as is evident in drawings in picture scrolls of the day. In one scroll especially, the *Honen Shōnin Eden*, numerous fences are depicted, including brushwood, bamboo branch, and wickerwork fences; that a Kenninji fence is also depicted is very interesting (Fig. 5).

The word *sode-gaki* (sleeve or wing fence) appeared during this period as well. A poem by Minister of the Left Hanazono found in the early fourteenth-century tanka collection *Fuboku Waka Shū* reads:

> Evidently the handiwork
> of an unrefined
> mountain dweller—
> open morning glories climbing
> on the brushwood *sode-gaki*

This is a very early mention of this kind of fence.

The next important stage in the development of bamboo fences in Japan took place during the Momoyama period (1568–1615), when the tea ceremony was refined by such masters as Sen no Rikyū. Tea-ceremony gardens were developed, and bamboo fences became a requisite fixture of these. Simple see-through fences, such as four-eyed fences, and subdued brushwood fences were used most. Eventually, wing fences were erected next to tea-ceremony houses as well (Fig. 6). These fences were

Fig. 5

Fig. 6

Fig. 7

Fig. 8

not made to serve as mere barriers, but were built to add the proper atmosphere to the garden and to fit the spirit of the tea ceremony itself. As such, they are a superb expression of Japanese aesthetics.

The next two and a half centuries, the Edo period, were a time of peace under the Tokugawa shogunate. Creativity in the development of bamboo fences for ordinary and tea-ceremony gardens flourished throughout the nation, giving rise to a large variety of fences. Bamboo fences were depicted in the paintings and illustrations of the day, and several simple drawings of them appeared in a tea-ceremony guide entitled *Kokon Sado Zensho*, published in 1694.

"Secret books" passing on the art of garden making were made during the period and included mentions of bamboo fences, and illustrated books contained drawings of them. One secret book on garden making, entitled *Tsukiyama Senshi Roku* (1797), lists thirteen fences, some of which are virtually unknown today. Two of particular interest are the Nanzenji fence and the Myōshinji (Myōshinji Temple) fence (*myōshinji-gaki*). Unfortunately, there is no information given about the former but that it is of bamboo. It is unclear whether this Nanzenji fence is the same as the original one (discussed in the main text of this book), but it is likely that it is not. The only words accompanying the Myōshinji fence entry are "bamboo branches." Although we can conjecture that it is one type of bamboo branch fence, we have absolutely no idea what it looked like. However, we do know that because of the close relationship between Zen and the tea ceremony, fences originally built around and named after Zen temples were constructed for tea-ceremony gardens as well.

Two works by Ritōken Akisato depict many bamboo fences: one, a collection on Kyoto gardens entitled *Miyako Rinsen Meishō Zue* (1799), and the other, a secret book on garden making entitled *Iwagumi Sonou Yae-gaki Den* (1827). In the latter Akisato presents illustrations and explanations of thirty-seven kinds of fences and fourteen kinds of gates (Figs. 7, 8). Although some of the fences listed are not seen today, many are, and this book is the earliest book known in which many of them are mentioned. Some fences in *Iwagumi Sonou Yae-gaki Den* that are described in more detail here are the Kenninji fence, the numazu fence, the teppō wing fence, the four-eyed fence, and the Ōtsu fence.

MATERIALS

One reason bamboo fence making developed to the extent it did in Japan is the large amount of good, suitable bamboo available. What kinds of bamboo were used in old times is not known, although we may speculate that the madake, hachiku, and medake varieties were among them. The bamboo most often mentioned in the early literature is kuretake bamboo, probably what we know today as hachiku, a sturdy variety. The early-fourteenth-century miscellany *Tsurezure-gusa* states: "Kuretake has fine leaves, and those of kawatake bamboo are broad. The one near the [Seiryōin Palace's] garden ditch is kawatake, the one growing in the area of the Jijuden [a mansion of the palace] is kuretake." Kawatake is now known as medake, which is one subvariety of shino, a fine bamboo.

The most suitable variety of bamboo for fence building, however, is madake. Madake, along with hachiku, was originally cultivated in Japan. (Perhaps it is because madake did not exist in China that bamboo fence construction never gained currency there.) It is very straight, its branches are hard, the interjoint space is long, and its wood is thin. Madake is strong and does not rot easily. It grows in varying thicknesses: stouter kinds are up to four inches in diameter, and finer kinds are a bit more than an inch thick (these latter are also called garadake).

One other important variety of bamboo, mōsōchiku, was not brought to Japan until about 250 years ago, from the Jiangnan region of China. It also became a popular material for bamboo fences, but because of the short interjoint space and thick wood of this variety, it is generally considered inferior to madake for fence construction. However, since its branches are quite pliant, it is a good material for bamboo branch fences.

CLASSIFICATION

There is a great variety of bamboo fences. Within any given type there are subvarieties, and different kinds may be combined into one fence. Their names are almost always based on temple names or on Japanese words from everyday life and often change with time. Thus, a classification of bamboo fences is very difficult; in fact, a perfect classification is probably impossible.

The following presents several easily understandable classification schemes. Below each group are varying numbers of examples. (The translated names of the fences are used; where these differ from the Japanese name, that name is given in parentheses.) Because the fences can be classified by different variables, there is naturally going to be some overlap. A few fences not included in this book are listed below for the reader's reference.

I. CLASSIFICATION BASED ON VISIBILITY

A. Screening fences (*shahei-gaki*; fences that cannot be seen through)
- Kenninji
- shimizu
- tokusa
- spicebush (*kuromoji*)
- bamboo branch (*takeho*)
- Katsura
- Ōtsu
- Numazu fences

B. See-through fences (*sukashi-gaki*)
- Four-eyed (*yotsume*)
- Kinkakuji
- Ryōanji
- stockade (*yarai*)
- Kōetsu
- Nison'in fences

Note: Fences can be made combining these two distinctions, for example, Kenninji and bamboo branch fences with lower parts that can be seen through (shita sukashi)*.*

II. CLASSIFICATION BASED ON USE

A. Enclosing fences (*kakoi-gaki*; usually screening fences, sometimes see-through fences)

B. Partitioning fences (*shikiri-gaki*; any kind of screening fence or see-through fence can be used)

III. CLASSIFICATION BASED ON HEIGHT

A. Fences of ordinary height (three feet or more high)
- Kenninji
- shimizu
- tokusa
- teppō
- Katsura
- bamboo screen (*misu*)
- raincoat (*mino*)
- Ōtsu
- Numazu
- four-eyed (*yotsume*) fences

B. Low fences (*ashimoto-gaki*, or "foot-level fences")
- Kinkakuji
- Ryōanji
- Nison'in
- nanako fences

*Note: Stockade (*yarai*) fences can be made either way, and four-eyed fences can be made low.*

IV. CLASSIFICATION BASED ON RELATIONSHIP TO BUILDINGS

    A. Fences made away from buildings

    B. Fences integrally related to buildings, e.g., wing fences (*sode-gaki*)

    C. Movable fences, e.g., *tsuitate* fences

V. CLASSIFICATION BASED ON THE MATERIALS USED FOR THE FRETS AND VERTICAL POLES (*KUMIKO* AND *TATEKO*)

    A. Bamboo trunks used
- Kenninji
- Ginkakuji
- shimizu
- tokusa
- bamboo screen (*misu*)
- Ōtsu
- four-eyed (*yotsume*)
- Kinkakuji
- Ryōanji
- Kōetsu fences

    B. Bamboo branches used
- Bamboo branch (*takeho*)
- Katsura
- raincoat (*mino*) fences

    C. Tree branches used
- spicebush (*kuromoji*)
- bush clover (*hagi*)
- brushwood (*shiba*)
- uguisu fences

    D. Tree bark used
- Japanese cypress bark (*hiwada*) fences

    E. Combinations of the above used
- Nanzenji and teppō fences

VI. CLASSIFICATION BASED ON THE ARRANGEMENT OF FRETS AND VERTICAL POLES (*KUMIKO* AND *TATEKO*)

    A. Fences using vertical poles (*tateko*)

      1. Using a single row of vertical frets
- Kenninji
- Ginkakuji
- shimizu
- shino
- Nanzenji
- zuiryū fences

      2. Using an alternating (back-front, back-front) arrangement of vertical poles (*teppō-zuke*)
- Four-eyed (*yotsume*)
- teppō fences

    B. Fences using frets (*kumiko*) horizontally arranged
- Bamboo screen (*misu*)
- Katsura fences

    C. Fences using frets (*kumiko*) diagonally arranged
- stockade (*yarai*)
- Ryōanji
- Kōetsu
- Numazu
- Kōrai fences

    D. Fences with woven fretwork
- Ōtsu
- Numazu
- wickerwork (*ajiro*) fences

E. Fences with layered fretwork
- bamboo branch (*takeho*)
- raincoat (*misu*)
- armor (*yoroi*) and
- shigure fences

F. Fences with fretwork attached in neat rows
- Japanese cypress bark (*hiwada*) and
- hishigi fences

VII. NOMENCLATURE CLASSIFICATIONS

A. Fences named after the material used (usually the material for the poles and frets, *tateko* and *kumiko*)
- shimizu fence: made of shimizudake bamboo
- spicebush (*kuromoji*) fence: made of branches of this plant
- bush clover (*hagi*) fence: made of branches of the Japanese bush clover
- bamboo branch (*takeho*) fence
- brushwood (*shiba*) fence: made of the branches of various trees
- shino fence: made of shino bamboo
- Japanese cypress bark (*hiwada*) fence
- hishigi fence: made of hishigidake bamboo

B. Fences named after places
- Ōtsu fence: after early fences made in Ōtsu, Shiga Prefecture (traditional derivation)
- Numazu fence: after fences built near Numazu, Shizuoka Prefecture
- Kōrai fence: because the fences have a Korean appearance ("Kōrai" is the name of an early Korean dynasty and an old Japanese word for the Korean peninsula)

C. Fences named after Buddhist temples (*-ji*) and other specific places
- Kenninji fence: first made at Kenninji in Kyoto (traditional)
- Daitokuji fence: first made at Daitokuji in Kyoto (traditional)
- Sōkokuji fence: first made at Sōkokuji in Kyoto (traditional)
- Chōfukuji fence: first made at Chōfukuji in Kyoto (traditional)
- Ginkakuji fence: first made at Jishōji (also called Ginkakuji) in Kyoto
- Nanzenji fence: first made at Nanzenji in Kyoto
- Kinkakuji fence: first made at Rokuonji (also called Kinkakuji) in Kyoto
- Ryōanji fence: first made at Ryōanji in Kyoto
- Nison'in fence: first made at Nison'in (a temple) in Kyoto
- Katsura fence: first made at the Katsura Detached Palace in Kyoto

D. Fences named after people
- Kōetsu fence: after Hon'ami Kōetsu (1558–1637), a Kyoto craftsman (The fence is also called a Kōetsuji fence, in which case it can be classified under the temple grouping, above.)
- Rikyū fence: after Sen no Rikyū (1521–91), a tea master
- Enshū fence: after Kobori Enshū (1579–1647), a tea master
- Sōwa fence: after Kanamori Sōwa (1758–1656), a tea master
- Narihira fence: after Ariwara-no Narihira (825–880), a poet
- Komachi fence: after Ono no Komachi, female poet of the Heian period (794–1185)

E. Fences whose names are based on their appearance, especially on other things they resemble
- tokusa fence: arrangement of fretwork (*tateko*) resembles tokusa, a kind of rush
- teppō fence: resembles vertical rows of rifle barrels (*teppō*)
- bamboo screen (*misu*) fence: resembles a bamboo screen
- raincoat (*mino*) fence: resembles an old straw raincoat
- four-eyed (*yotsume*) fence: rows are arranged in a way that leaves four open spaces
- stockade (*yarai*) fence: diagonal arrangement of fretwork (*kumiko*) resembles a bamboo stockade
- nanako fence: resembles the ancient nanako pattern
- uguisu fence: the top of the fence looks like an inviting place for uguisu (Japanese bush warblers) to nest
- tea whisk (*chasen*) fence: fence top resembles tea whisks
- torch (*taimatsu*) fence: vertical frets (*tateko*) are arranged into torch shapes
- armor (*yoroi*) fence: resembles ancient armor
- wickerwork (*ajiro*) fence: resembles a wickerwork pattern
- fusuma fence: resembles a sliding door (*fusuma*)
- nozoki fence: has windows that can be looked through (*nozoku*, in compound words, *nozoki*)
- three-tiered (*sandan*) fence: has three vertical levels
- tachiai fence: because of the alternating arrangement of two different types of vertical frets in the front of the fence
- tsuitate fence: resembles a small partitioning screen (*tsuitate*) used in a room
- folding screen (*byōbu*) fence: resembles a folding screen used in a room
- gagyū fence: resembles a cow (*gyō*) lying down (*ga*).
- aboshi fence: resembles the pattern made by nets (*a*, or *ami*) drying (*-boshi*, from *hoshi*) at the seaside
- moji fence: support poles (*oshibuchi*) are arranged to look like characters or letters (*moji*).

GLOSSARY

Note: A number of words not used in the main text are included below for reference.

ashimoto-gaki (foot-level fence): a low-lying fence, about knee height or less. Common *ashimoto-gaki* are the Kinkakuji, Ryōanji, Nison'in, and nanako fences; four-eyed fences are also made as *ashimoto-gaki*.

beading: see *tamabuchi*

bundled tateko: see *tateko*

-dake: the suffix form of *take*, bamboo

dōbuchi: translated here as "horizontal frame pole"; a main horizontal supporting piece extending between the posts of a fence. In screening fences, the *dōbuchi* are usually hidden, so they can be of ordinary pieces of wood. In see-through fences, however, they are usually visible, so round bamboo is often used. Some fences are made without *dōbuchi*.

fukiyose: an arrangement in which two long pieces of bamboo, such as horizontal support poles, are brought nearer to each other than usual (but do not touch).

furedome: a long thin piece of round or split bamboo attached horizontally near the top of a fence not having beading, to hold the vertical poles in place. Used for screening fences, notably bamboo branch fences and brushwood fences.

fushidome: the method of cutting bamboo poles just above the joints (*fushi*). This method results in stronger pieces and hides their inside (since there is a flat "plate" at the joints), thus making them more attractive and resistant to rain.

-gaki: the suffix form of *kaki*, fence

garadake: a name used among landscape architects for thin madake bamboo. Garadake is used mainly for four-eyed fences.

gyō: see *shin-gyō-sō*

hachiku: a sturdy species of bamboo

hashira, -bashira: a post. The main, sturdy posts of a bamboo fence, usually the two posts at the ends, or the single post of a wing fence, are called *oya-bashira* (or *tome-bashira* or *chikara-bashira*). Relatively narrow inner posts of a long fence, spaced about six feet apart, are called *ma-bashira*. *Maki-bashira* are log posts wound with fine round or split bamboo, bamboo branches, spicebush, or bush clover and are used mainly for Kōetsu, teppō, and wing fences.

hishigi-dake: a stalk of bamboo that has been crushed, producing numerous vertical cracks, and then "unfolded" into a more or less flat piece. A handsome fence, called a *hishigi-gaki*, is made when such pieces are arranged vertically so that their joints line up.

honka: translated here as "the original fence"; the first fence of a given type. The originals of most bamboo fence types have not been determined; some of those which have known *honka* are the Kinkakuji, Ryōanji, and Kōetsu fences.

horizontal frame pole: see *dōbuchi*

kaki, -gaki: a fence

Karage shuho (karage method): a joining method using rope, but not involving tying. A single piece of rope is wound around a vertical support pole. Used mostly with the four-eyed fence, the *karage* method has such variations as *yotsume karage* and *kaizuka karage*.

knots: see *nawa musubi*

kumiko: translated here as "frets" or "fretwork"; diagonal or horizontal branches of wood or pieces of bamboo serving a decorative (as opposed to supporting) function in bamboo fences. Among screening fences, *kumiko* are found mainly in the Numazu fence and bamboo screen fence; among see-through fences, *kumiko* are used mainly in the stockade fence, Ryōanji fence, and Kōetsu fence.

kurochiku: a black variety of hachiku bamboo

kuroho: the black branches of kurochiku bamboo. Used mainly in the Kantō area and often for raincoat fences, kuroho give a fence a subdued atmosphere.

madake: a common species of bamboo

midare shuhō (midare method): a method of purposely making the lengths of the vertical support poles different so that the top of a fence is uneven. When this method is used for Kenninji fences and four-eyed fences, the result is a *sō* fence.

mōsōchiku: a species of thick-stemmed bamboo

mume ita: a piece of wood running between two posts, connected to the posts near ground level. Vertical support poles resting on the *mume ita* do not rot as easily as they otherwise would.

nawa musubi: rope tying, the main method used for holding bamboo fences together. The basic form is *ibo musubi*, also called simply *ibo*, or *yuibo*, *otoko musubi* (man's knot), or *yotsume no otoko musubi*; as the last variant indicates, one of the main uses of *ibo musubi* is for four-eyed fences (*yotsume-gaki*). A *kari musubi* is an *ibo musubi* purposely tied to come apart when one of the ends is pulled. Tied rope also often serves an important decorative purpose (*kazari musubi*), the most common method being beading *musubi* (which has several variants) used to attach the beading to the top of the fence. Ropes were made of the hemp palm (hardly available today) and the coconut palm, which may be dyed black (*somenawa*). *Somenawa* are well soaked in water to soften them before using. A sturdier rope, called *warabi nawa*, is made from fibers of bracken (*warabi*). *Warabi nawa* is rather expensive, so its use today is limited mostly to small wing fences.

oshibuchi: translated here as "support pole"; pieces of bamboo placed tightly over vertical poles and frets to hold them in place. Stout bamboo split in half is usually used, although fine round-bamboo *oshibuchi* are sometimes seen. *Oshibuchi* are usually placed horizontally, but in such fences as the Katsura fence and the bamboo screen fence, they are arranged vertically. The arrangement of the *oshibuchi* influence greatly the overall beauty of a fence.

post: see *hashira*

sarashidake: a processed form of madake or hachiku bamboo. Thin pieces of bamboo are heated over a flame to remove the oils (today, chemicals are often used instead of fire), and the bamboo is straightened. Used often for the frets of bamboo screen fences.

sashi ishi: small, flat stones upon which a bamboo fence rests; the stones are placed on the ground to support the fence, since the part of the posts in the ground will eventually rot.

screening fence: see *shahei-gaki*

see-through fence: see *sukashi-gaki*

shahei-gaki: a bamboo fence that cannot be seen through. A typical variety is the Kenninji fence.

shimizudake: polished, straightened shino bamboo from which the oils have been removed. Pieces are of a set length and used to make shimizu fences.

shin-gyō-sō: refers to three forms of style in many fields. *Shin* is the most formal, proper style; the *sō* form is a freer style; and the *gyō* is a form between the two. In bamboo fence-making, the Kenninji fence, four-eyed fence, and others exhibit the three forms.

shino, shinodake: a small, thin kind of bamboo; varieties include yadake, medake, and hakonedake.

shinobi no take: thin split bamboo used to hold bamboo branches in place during the process of constructing a bamboo fence; also called simply *shinobi*. Sometimes the *shinobi no take* are left as part of the fence if they are invisible; in some processes they are removed entirely before the fence is completed.

shiori: the process of severely bending bamboo or wood.

shiroho: light-colored bamboo branches of such varieties as mōsōchiku, hachiku, and madake bamboo.

sleeve fence: see *sode-gaki*

sō: see *shin-gyō-sō*

sode-gaki (lit. "sleeve fence"): translated here as "wing fence"; any of many varieties of a small bamboo fence with a single post standing against a building, used mostly for decorative purposes; so named because it resembles the sleeve (*sode*) of a kimono. Screening fences are the more common type (often seen alongside gates), but see-through fences also exist.

sukashi-gaki: a bamboo fence that can be seen through, used widely for fences serving as inner partitions in gardens. The Kōetsu fence and four-eyed fence are typical examples.

support pole: see *oshibuchi*

take, -dake: bamboo

takegaki: a bamboo fence

takeho: bamboo branches

tamabuchi: translated here as "beading"; bamboo or other molding placed along the top of a fence as decoration and to protect the fence from rain. *Tamabuchi* is made of split bamboo, bundled bamboo, or bush clover (*hagi*) branches, and bamboo branches covered with fine split bamboo. When split bamboo is used, horizontal support poles may be attached to both sides of the top of the fence and a *kasadake* placed on the top; the entire arrangement is called a *tamabuchi*. Fences of a given type may have varieties with or without *tamabuchi*.

tateko: translated here as "vertical pole" or "vertical piece"; a vertical *kumiko* (fret) of round bamboo, split bamboo, bamboo branches, wood, etc. *Maki tateko* are branches of bamboo, spicebush (*kuromoji*), bush clover (*hagi*), etc., wound and bundled together; they are used mainly for teppō fences.

tying methods: see *nawa musubi*

vertical piece, pole: see *tateko*

warima: the spacing of horizontal elements with respect to the overall height of a fence, and of vertical elements with respect to the width between posts.

wing fence: see *sode-gaki*

yama wari-dake: standard (six-foot) lengths of stout split bamboo, used mainly for the vertical support poles of Kenninji fences.

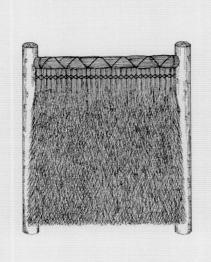